The LIFE CYCLES Library™

The LIFE CYCLE of a SNAIL

Andrew Hipp

Photography by Dwight Kuhn

The Rosen Publishing Group's
PowerKids Press
New York

For my family—Andrew Hipp
For Jarrett—Dwight Kuhn

Published in 2002 by The Rosen Publishing Group, Inc.
29 East 21st Street, New York, NY 10010

First Edition

Book Design: Michael Caroleo and Michael de Guzman
Project Editor: Emily Raabe

All Photos © Dwight Kuhn.

The author gratefully acknowledges David Lewis, who commented on an early draft of this book.

Hipp, Andrew.
The life cycle of a snail / Andrew Hipp.
    p. cm. — (The life cycles library)
Includes bibliographical references (p.   ).
    ISBN: 978082958719

1. Snails—Life cycles—Juvenile literature. [1. Snails.] I. Title.
QL430.4 .H57 2002
593'.3—dc21   2001000279

Manufactured in the United States of America

Contents

Looking for Snails

Snails live in deserts, on mountains, in forests, and in oceans. Some snails breathe air using lungs, while other snails live only in the water. If you want to find a snail, look under old logs in a park or in your backyard. Poke around the fallen leaves in a forest. The first thing you might see is the snail's shell. Most land snails also have four **antennae**, called **tentacles**, with eyes at the tips of the upper pair. Snails can retract, or pull back, their tentacles to protect their eyes.

◀ *If you visit a pond or creek with your parents, sooner or later you will come across a snail, or at least an old snail shell.*

Mating

Many snails are **hermaphrodites**. This means that the snail has both male and female parts. Snails crawl next to each other to mate. Each snail squirts **sperm** into the shell of the other snail. This sperm will be used to **fertilize** the eggs of the other snail. Snails can store sperm for about 100 days, using it to fertilize many batches of eggs. Some snails can store sperm for as long as a year!

The bubbles in this photo are produced by the mating snails. ▶

Laying Eggs

Snails lay their eggs from eight days to several weeks after mating. Some snails dig a nest hole in the soil. Other snails lay eggs in holes that they find. Land snails look for damp places to lay their eggs. They lay eggs with shells that keep the baby snails inside from drying out. Freshwater snails don't need to worry about their eggs drying out. They lay eggs in long, gooey strings under water or on water plants. Some snails keep their eggs inside their bodies until they hatch.

◀ *Snail eggs have small yolks to feed the growing baby snails. Bird eggs have much bigger yolks than do snail eggs.*

Hatching

One week or so after an egg is laid, the baby snail begins to creep around on the inside of the eggshell. It scratches at the shell with a sharp organ in its mouth, called a **radula**. The snail hatches from one to six weeks after the egg was laid. Many snail eggs never hatch. Animals may eat the eggs. Eggs also dry out, killing the babies inside. If the eggs do hatch, both land and freshwater snails may stay in their nests for one week or more afterward.

When snails are first born, you can see right through their shells. ▶

Eating

Snails eat mostly plants and dead animals. Sometimes they eat insects. Some snails eat so many plants that they are pests to gardeners and farmers. Most snails, however, like dead, rotting leaves more than living plants. They will eat even damp paper or cardboard, which is made from dead trees. Snails are important **scavengers**. This means that dead things are a large part of their diets. Scavengers are important because they help to turn dead animals and plants into soil.

◀ *Hungry snails can damage garden plants or crops.*

Growing Shells

Every kind of snail has a shell of a different shape, size, or color. A snail's shell is strong and helps to protect it from drying out. It also protects the snail from **predators**. Fish, turtles, diving beetles, firefly larvae, leeches, mink, raccoons, ducks, crayfish, and birds all eat snails. Even some snails eat snails! Snails avoid some predators by crawling away from them. A big shell also helps to make a snail much harder to eat.

The ridges on a snail's shell form as it grows. Each ridge is a new layer of shell material. ▶

The Radula

A snail has something in its mouth called a radula. The radula looks like a tongue and is covered with small, razor-sharp teeth. A snail can have more than 25,000 teeth on its radula! The snail uses its radula to scrape food off leaves and other surfaces. It also uses its radula to bite off pieces of food. One kind of snail eats mainly earthworms. This snail stabs an earthworm with its long, pointed radula teeth and drags the worm into its mouth.

◄ *This snail will use its sharp radula to scrape food off of leaves, tree bark, and other surfaces.*

17

Winter

When winter or dry weather comes, land snails **hibernate** under leaves, logs, or soil. Once it finds a safe place, a snail pulls its entire body into its shell. It closes the opening of its shell with several layers of **calcium** or **mucus**. These layers keep the snail from drying out. Some snails that live in dry areas can hibernate for years, waiting for rain. They use almost no energy and eat nothing. When spring or a wet season comes, snails awaken again.

Land snails hibernate in protected places, such as this flower pot. Freshwater snails spend winter beneath the ice, moving slowly or not at all. ▶

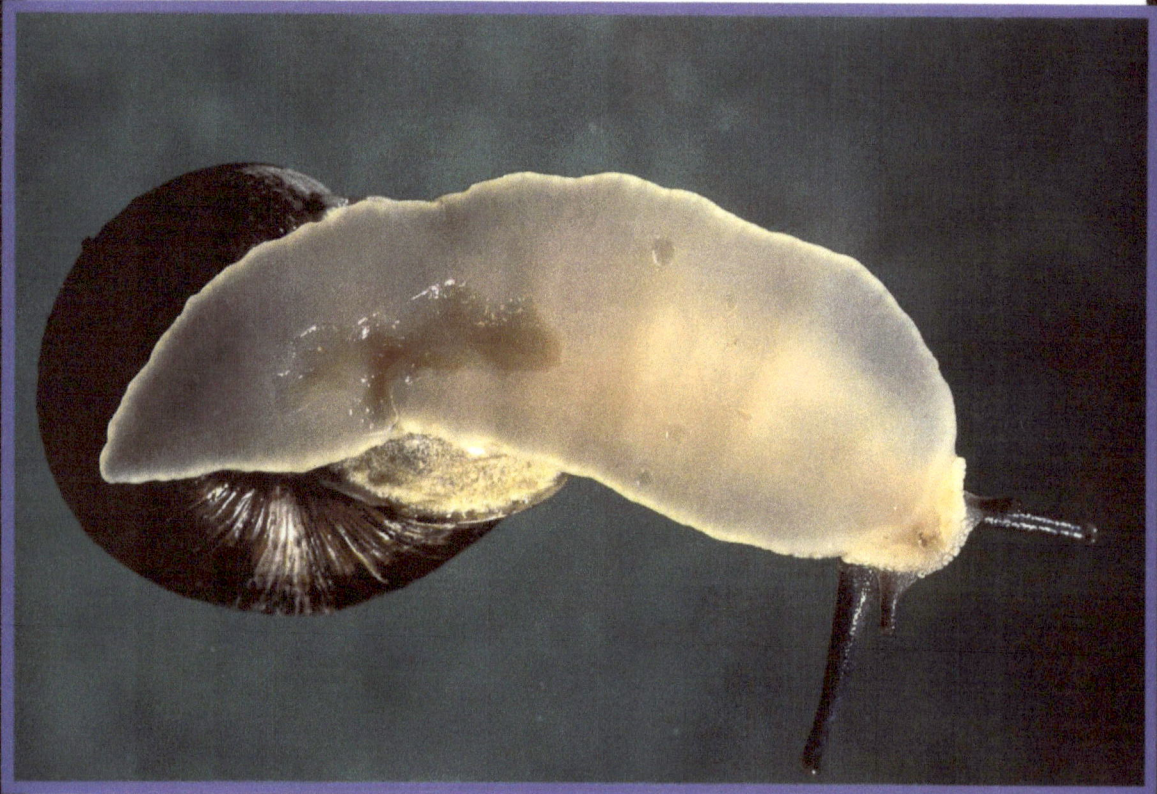

Shells All Around

Snails are part of a group of animals called **mollusks**. Clams, oysters, slugs, squid, and octopi are also mollusks. Most mollusks have shells and strong feet, like snails. Many have radulae as well. When you find shells at the beach or at the edge of a pond, you are finding what is left of dead mollusks. Snail shells often fall to the bottoms of ponds and lakes, where they become important parts of the soil. Some soils near ponds or lakes are crunchy with shells.

◄ *Most of what you see in this photo is the snail's foot. Its radula is the tiny dark spot visible in its mouth.*

Taking Care of Snails

You can raise freshwater snails at home in a big jar or in an aquarium. Snails need special water plants to eat. You can get plants and snails at pet stores that sell aquariums. You can also get them from a pond. Bring an adult, and a net for scooping snails. Your snails will need sand, plants, and fresh water. Some tap water has a gas in it, called chlorine, that kills snails, so if you use tap water for your snails, let it sit for 24 hours first to let any chlorine out of the water. With clean water and light for the plants, your snails will be as happy as can be!

Glossary

antennae (an-TEH-nee) Thin, rodlike organs on the head of insects, lobsters, and snails, used for smelling or feeling.

calcium (KAL-see-um) A mineral that is an important part of bones, teeth, shells, and egg shells.

fertilize (FUR-til-eyez) Introducing male reproductive cells into the female to make babies.

hermaphrodites (her-MA-froh-dyts) Animals that have both male and female reproductive organs.

hibernate (HY-bur-nayt) To spend time in a state of rest, without eating or moving.

mollusks (MAH-lusks) Animals without backbones, usually with shells for protection.

mucus (MEW-kus) Thick, sticky liquid made in the body.

predators (PREH-duh-terz) Animals that kill other animals for food.

radula (RA-jeh-leh) A tongue-like organ in the mouth of a snail.

scavengers (SKA-ven-jurz) Animals that feed on dead animals.

sperm (SPURM) Fluid that male animals produce when mating.

tentacles (TEN-tuh-kuhls) The name for antennae on snails' heads.

Index

Web Sites

To learn more about snails, check out these Web sites:
www.EnchantedLearning.com/subjects/invertebrates/mollusk/
    gastropod/Snailprintout.shtml
www.kiddyhouse.com/Snails/

www.ingramcontent.com/pod-product-compliance
Lightning Source LLC
Chambersburg PA
CBHW050913210326
41597CB00002B/106